Star Gazing Journal

John Philip Hunter

Published by FunTastic Press, an imprint of SRS Internet Publishing, 2024.

www.FunTasticPress.com

ISBN: 978-1-936456-66-6

Printed in the United States of America.

This Star Gazing Journal Belongs To:

____

Observation Dates From: ____________ to: ____________

Introduction

Throughout history, humanity's fascination with astronomy has endured, evolving from ancient wonder to modern exploration. Today, our curiosity extends beyond observation to active investigation of the stars and planets. While science-fiction programs dominate television screens, real-life space exploration endeavors are thriving once again, captivating both individuals and nations alike. In this era of boundless information accessible via the Internet, the allure of space travel is coupled with unprecedented accessibility to astronomical knowledge.

However, amidst this excitement, it remains crucial to approach stargazing with the mindset of a diligent astronomer. Embracing the scientific method, meticulous note-taking becomes an integral part of the experience. In the past, my own forays into star gazing involved scribbling observations on scraps of paper or in modest spiral notebooks, occasionally upgrading to more elaborate bound journals. It is in this spirit that I present to you the *Star Gazing Journal*, thoughtfully designed to streamline your data entry process with pre-printed information fields, facilitating swift and efficient documentation.

When utilizing the journal, remember to record not only the time and weather conditions but also your precise location, as environmental factors such as cloud cover and lunar phases significantly impact observational quality. Ample space is provided for sketching special star configurations and noting celestial phenomena, ensuring a comprehensive record of your experiences. -- JPH

Terms & Tips for Observational Recording

Below, you'll find some key points to optimize your journaling experience and extract the utmost value from your observations.

Aperture : This refers to the diameter of the lens or mirror which gathers the light. Usually this is noted because each telescope may be different. Aperture gives you information as to the brightness of viewing, thus seeing more detail.

FOV: The field of view (also field of vision, abbreviated FOV or instantaneous field of view, abbreviated IFOV) is the extent of the observable world that is seen at any given moment.

EP: Refers to the eye piece of your telescope.

Transparency: In astronomy the transparency of the Earth's atmosphere is important because it determines the ability of astronomers to observe from Earth rather than from space. The Earth's atmosphere transmits mainly visible light, infrared light, radio, microwave and millimeter wavelength radiation. Seeing is largely an atmospheric issue, but telescope optics plays a part. The end result is that refractors tend to provide steadier images than reflectors.

Astronomical Seeing: The term Astronomical Seeing is unique to astronomy in that it refers to the effect the atmosphere has on what you are viewing. Remember, that Earth's atmosphere is likened to fluid. Thus, at times it can be clear and other times murky or cloudy. In your journal note the actual weather conditions and consider the transparency as a separate point. These weather notes go under the heading of Conditions.

Conditions: The weather is one of the key points to your viewing. A hot day on Earth can make for fluctuations in the atmosphere and cause blurred views. Even the area you are star gazing can have air turbulence if the day has been very hot.

Latitude & Longitude: Here you can note the celestial coordinates of latitude and longitude of your location if you wish.

Lunar Phase: Some sky watchers like to note the phase of the moon for that date.

Using a journal for star gazing and sky watching is more than scientific… it is fun! So, get out there and start recording your observations! Who knows? Maybe you'll discover the next new comet or asteroid!

NAVIGATIONAL STAR CHART

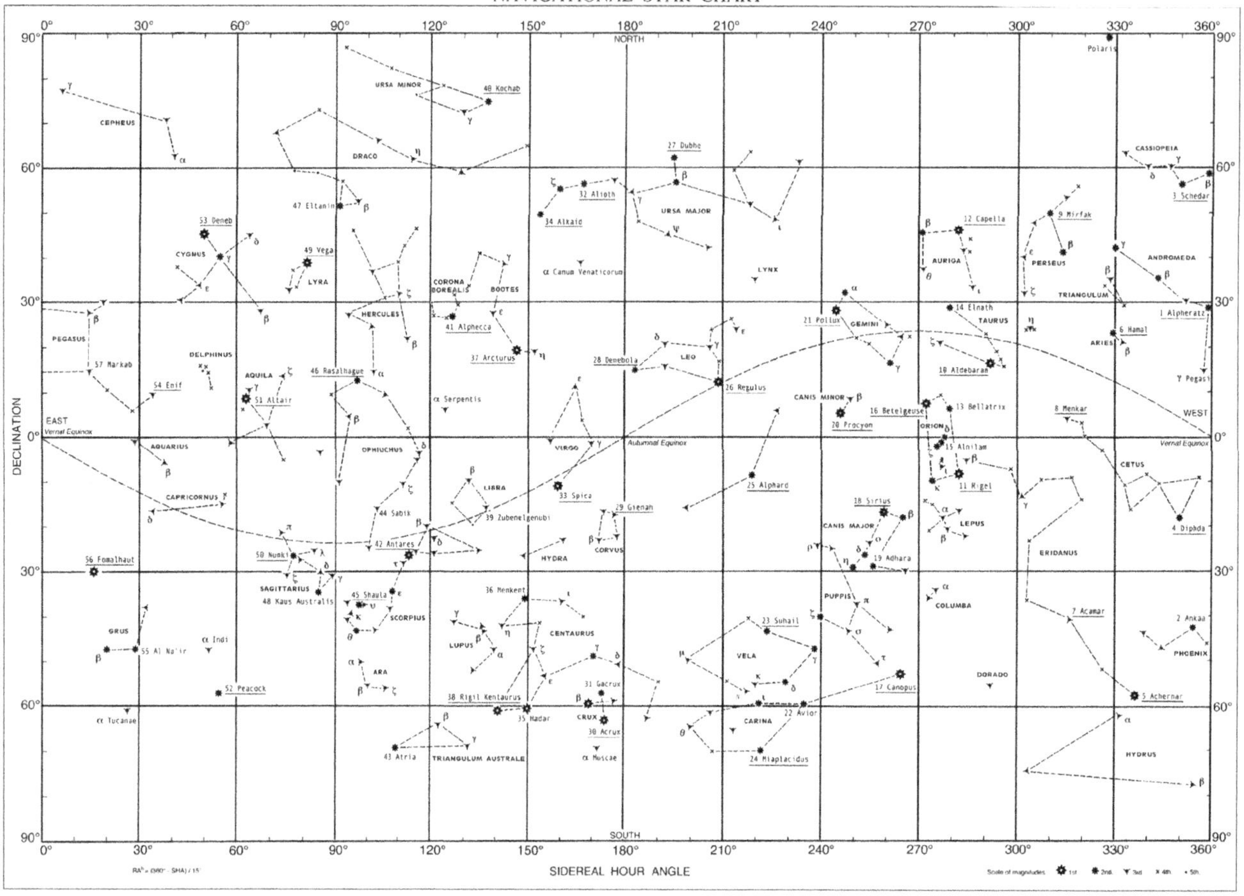

Date: ______________________ Time: ______________________

Location: ______________________ Lunar Phase: ______________________

Longitude: ______________________ Latitude: ______________________

Object(s) Viewed: ______________________

Equipment:

Filter	Magnification	EP	Aperture	FOV

Conditions	Seeing	Finder

Transparency: ______________________

Notes:

Date:____________________ Time: ____________________

Location:____________________ Lunar Phase: ____________________

Longitude: ____________________ Latitude: ____________________

Object(s) Viewed: ____

Equipment:

Filter	Magnification	EP	Aperture	FOV

Conditions	Seeing	Finder

Transparency: ____

Notes:

Date: ______________________ Time: ______________________

Location: ______________________ Lunar Phase: ______________________

Longitude: ______________________ Latitude: ______________________

Object(s) Viewed: ____

Equipment:

Filter	Magnification	EP	Aperture	FOV

Conditions	Seeing	Finder

Transparency: ____

Notes:

Date: ____________________ Time: ____________________

Location: ____________________ Lunar Phase: ____________________

Longitude: ____________________ Latitude: ____________________

Object(s) Viewed: ____

Equipment:

Filter	Magnification	EP	Aperture	FOV

Conditions	Seeing	Finder

Transparency: ____

Notes:

Date: ______________________ Time: ______________________

Location: ______________________ Lunar Phase: ______________________

Longitude: ______________________ Latitude: ______________________

Object(s) Viewed: ____

Equipment:

Filter	Magnification	EP	Aperture	FOV

Conditions	Seeing	Finder

Transparency: ____

Notes:

Date:______________________________ Time: __________________________________

Location:___________________________ Lunar Phase: __________________________

Longitude: _________________________ Latitude: _______________________________

Object(s) Viewed: ____

Equipment:

Filter	Magnification	EP	Aperture	FOV

Conditions	Seeing	Finder

Transparency: ____

Notes:

Date:______________________ Time: ______________________

Location:______________________ Lunar Phase: ______________________

Longitude: ______________________ Latitude: ______________________

Object(s) Viewed: ____

Equipment:

Filter	Magnification	EP	Aperture	FOV

Conditions	Seeing	Finder

Transparency: ____

Notes:

Date: ____________________ Time: ____________________

Location: ____________________ Lunar Phase: ____________________

Longitude: ____________________ Latitude: ____________________

Object(s) Viewed: ____

Equipment:

Filter	Magnification	EP	Aperture	FOV

Conditions	Seeing	Finder

Transparency: ____

Notes:

Date: ______________________________ Time: _________________________________

Location: ____________________________ Lunar Phase: __________________________

Longitude: __________________________ Latitude: _______________________________

Object(s) Viewed: ____

Equipment:

Filter	Magnification	EP	Aperture	FOV

Conditions	Seeing	Finder

Transparency: ____

Notes:

Date:______________________ Time: ______________________

Location:____________________ Lunar Phase: __________________

Longitude: __________________ Latitude: ____________________

Object(s) Viewed: ____

Equipment:

Filter	Magnification	EP	Aperture	FOV

Conditions	Seeing	Finder

Transparency: ____

Notes:

Date:______________________________ Time: _________________________________

Location:___________________________ Lunar Phase: __________________________

Longitude: _________________________ Latitude: ______________________________

Object(s) Viewed: ____

Equipment:

Filter	Magnification	EP	Aperture	FOV

Conditions	Seeing	Finder

Transparency: ____

Notes:

Date:_______________________________ Time: _________________________________

Location:___________________________ Lunar Phase: ___________________________

Longitude: __________________________ Latitude: ______________________________

Object(s) Viewed: ____

Equipment:

Filter	Magnification	EP	Aperture	FOV

Conditions	Seeing	Finder

Transparency: ____

Notes:

Date: ______________________ Time: ______________________

Location: ______________________ Lunar Phase: ______________________

Longitude: ______________________ Latitude: ______________________

Object(s) Viewed: ____

Equipment:

Filter	Magnification	EP	Aperture	FOV

Conditions	Seeing	Finder

Transparency: ____

Notes:

Date: ______________________ Time: ______________________

Location: ______________________ Lunar Phase: ______________________

Longitude: ______________________ Latitude: ______________________

Object(s) Viewed: ____

Equipment:

Filter	Magnification	EP	Aperture	FOV

Conditions	Seeing	Finder

Transparency: ____

Notes:

Date:______________________________ Time: ________________________________

Location:___________________________ Lunar Phase: __________________________

Longitude: _________________________ Latitude: _____________________________

Object(s) Viewed: ____

Equipment:

Filter	Magnification	EP	Aperture	FOV

Conditions	Seeing	Finder

Transparency: ____

Notes:

Date: ______________________ Time: ______________________

Location: ______________________ Lunar Phase: ______________________

Longitude: ______________________ Latitude: ______________________

Object(s) Viewed: ______________________

Equipment:

Filter	Magnification	EP	Aperture	FOV

Conditions	Seeing	Finder

Transparency: ______________________

Notes:

Date:____________________________ Time: ______________________________

Location:________________________ Lunar Phase: ________________________

Longitude: ______________________ Latitude: ____________________________

Object(s) Viewed: ____

Equipment:

Filter	Magnification	EP	Aperture	FOV

Conditions	Seeing	Finder

Transparency: ____

Notes:

Date:______________________________ Time: ______________________________

Location:______________________________ Lunar Phase: ______________________________

Longitude: ______________________________ Latitude: ______________________________

Object(s) Viewed: ______________________________

Equipment:

Filter	Magnification	EP	Aperture	FOV

Conditions	Seeing	Finder

Transparency: ______________________________

Notes:

Date:______________________________ Time: _________________________________

Location:__________________________ Lunar Phase: __________________________

Longitude: ________________________ Latitude: ______________________________

Object(s) Viewed: ____

Equipment:

Filter	Magnification	EP	Aperture	FOV

Conditions	Seeing	Finder

Transparency: ____

Notes:

Date: ______________________________ Time: ______________________________

Location: ___________________________ Lunar Phase: ________________________

Longitude: __________________________ Latitude: ___________________________

Object(s) Viewed: ____

Equipment:

Filter	Magnification	EP	Aperture	FOV

Conditions	Seeing	Finder

Transparency: ____

Notes:

Date:______________________________ Time: _________________________________
Location:__________________________ Lunar Phase: ___________________________
Longitude: ________________________ Latitude: ______________________________
Object(s) Viewed: ____

Equipment:

Filter	Magnification	EP	Aperture	FOV

Conditions	Seeing	Finder

Transparency: ____

Notes:

Date:____________________ Time: ____________________

Location:____________________ Lunar Phase: ____________________

Longitude: ____________________ Latitude: ____________________

Object(s) Viewed: ____

Equipment:

Filter	Magnification	EP	Aperture	FOV

Conditions	Seeing	Finder

Transparency: ____

Notes:

Date:________________________ Time: ________________________

Location:____________________ Lunar Phase: ___________________

Longitude: __________________ Latitude: ______________________

Object(s) Viewed: ____

Equipment:

Filter	Magnification	EP	Aperture	FOV

Conditions	Seeing	Finder

Transparency: ____

Notes:

Date: ______________________________ Time: ______________________________

Location: ____________________________ Lunar Phase: ________________________

Longitude: ___________________________ Latitude: ___________________________

Object(s) Viewed: ____

Equipment:

Filter	Magnification	EP	Aperture	FOV

Conditions	Seeing	Finder

Transparency: ____

Notes:

Date: ____________________ Time: ____________________

Location: ____________________ Lunar Phase: ____________________

Longitude: ____________________ Latitude: ____________________

Object(s) Viewed: ____________________

Equipment:

Filter	Magnification	EP	Aperture	FOV

Conditions	Seeing	Finder

Transparency: ____________________

Notes:

Date: ______________________ Time: ______________________

Location: ______________________ Lunar Phase: ______________________

Longitude: ______________________ Latitude: ______________________

Object(s) Viewed: ____

Equipment:

Filter	Magnification	EP	Aperture	FOV

Conditions	Seeing	Finder

Transparency: ____

Notes:

Date:______________________ Time: ______________________

Location:______________________ Lunar Phase: ______________________

Longitude: ______________________ Latitude: ______________________

Object(s) Viewed: ____

Equipment:

Filter	Magnification	EP	Aperture	FOV

Conditions	Seeing	Finder

Transparency: ____

Notes:

Date: ____________________ Time: ____________________

Location: ____________________ Lunar Phase: ____________________

Longitude: ____________________ Latitude: ____________________

Object(s) Viewed: ____________________

Equipment:

Filter	Magnification	EP	Aperture	FOV

Conditions	Seeing	Finder

Transparency: ____________________

Notes:

Date: ______________________ Time: ______________________

Location: ______________________ Lunar Phase: ______________________

Longitude: ______________________ Latitude: ______________________

Object(s) Viewed: ____

Equipment:

Filter	Magnification	EP	Aperture	FOV

Conditions	Seeing	Finder

Transparency: ____

Notes:

Date:______________________________ Time: ______________________________

Location:______________________________ Lunar Phase: ______________________________

Longitude: ______________________________ Latitude: ______________________________

Object(s) Viewed: ____

____

Equipment:

Filter	Magnification	EP	Aperture	FOV

Conditions	Seeing	Finder

Transparency: ____

Notes:

Date:____________________ Time: ____________________

Location:____________________ Lunar Phase: ____________________

Longitude: ____________________ Latitude: ____________________

Object(s) Viewed: ____

Equipment:

Filter	Magnification	EP	Aperture	FOV

Conditions	Seeing	Finder

Transparency: ____

Notes:

Date: ______________________ Time: ______________________

Location: ______________________ Lunar Phase: ______________________

Longitude: ______________________ Latitude: ______________________

Object(s) Viewed: ____

Equipment:

Filter	Magnification	EP	Aperture	FOV

Conditions	Seeing	Finder

Transparency: ____

Notes:

Date: ____________________ Time: ____________________

Location: ____________________ Lunar Phase: ____________________

Longitude: ____________________ Latitude: ____________________

Object(s) Viewed: ____

Equipment:

Filter	Magnification	EP	Aperture	FOV

Conditions	Seeing	Finder

Transparency: ____

Notes:

Date: ____________________ Time: ____________________

Location: ____________________ Lunar Phase: ____________________

Longitude: ____________________ Latitude: ____________________

Object(s) Viewed: ____

Equipment:

Filter	Magnification	EP	Aperture	FOV

Conditions	Seeing	Finder

Transparency: ____

	Notes:

Date: ______________________ Time: ______________________

Location: ______________________ Lunar Phase: ______________________

Longitude: ______________________ Latitude: ______________________

Object(s) Viewed: ____

____

Equipment:

Filter	Magnification	EP	Aperture	FOV

Conditions	Seeing	Finder

Transparency: ____

Notes:

Date: ______________________________ Time: ______________________________

Location: ___________________________ Lunar Phase: _______________________

Longitude: __________________________ Latitude: __________________________

Object(s) Viewed: ____

____

Equipment:

Filter	Magnification	EP	Aperture	FOV

Conditions	Seeing	Finder

Transparency: ____

Notes:

Date:______________________________ Time: _________________________________

Location:___________________________ Lunar Phase: __________________________

Longitude: _________________________ Latitude: ______________________________

Object(s) Viewed: ____

Equipment:

Filter	Magnification	EP	Aperture	FOV

Conditions	Seeing	Finder

Transparency: ____

Notes:

Date:____________________ Time: ____________________

Location:____________________ Lunar Phase: ____________________

Longitude: ____________________ Latitude: ____________________

Object(s) Viewed: ____________________

Equipment:

Filter	Magnification	EP	Aperture	FOV

Conditions	Seeing	Finder

Transparency: ____________________

Notes:

Date:____________________________ Time: _______________________________

Location:_________________________ Lunar Phase: ________________________

Longitude: _______________________ Latitude: ___________________________

Object(s) Viewed: ____

Equipment:

Filter	Magnification	EP	Aperture	FOV

Conditions	Seeing	Finder

Transparency: ____

Notes:

Date: ____________________ Time: ____________________

Location: ____________________ Lunar Phase: ____________________

Longitude: ____________________ Latitude: ____________________

Object(s) Viewed: ____

Equipment:

Filter	Magnification	EP	Aperture	FOV

Conditions	Seeing	Finder

Transparency: ____

Notes:

Date:______________________ Time: ______________________

Location:______________________ Lunar Phase: ______________________

Longitude: ______________________ Latitude: ______________________

Object(s) Viewed: ____

Equipment:

Filter	Magnification	EP	Aperture	FOV

Conditions	Seeing	Finder

Transparency: ____

Notes:

Date: ______________________ Time: ______________________
Location: ______________________ Lunar Phase: ______________________
Longitude: ______________________ Latitude: ______________________
Object(s) Viewed: ______________________

Equipment:

Filter	Magnification	EP	Aperture	FOV

Conditions	Seeing	Finder

Transparency: ______________________

Notes:

Date:______________________________ Time: ________________________________

Location:__________________________ Lunar Phase: __________________________

Longitude: ________________________ Latitude: _____________________________

Object(s) Viewed: ____

Equipment:

Filter	Magnification	EP	Aperture	FOV

Conditions	Seeing	Finder

Transparency: ____

Notes:

Date: ____________ Time: ____________

Location: ____________ Lunar Phase: ____________

Longitude: ____________ Latitude: ____________

Object(s) Viewed: ____________

Equipment:

Filter	Magnification	EP	Aperture	FOV

Conditions	Seeing	Finder

Transparency: ____________

Notes:

Date:______________________________ Time: ______________________________

Location:__________________________ Lunar Phase: _______________________

Longitude: ________________________ Latitude: ___________________________

Object(s) Viewed: ____

____

Equipment:

Filter	Magnification	EP	Aperture	FOV

Conditions	Seeing	Finder

Transparency: ____

Notes:

Date:____________________ Time: ____________________

Location:____________________ Lunar Phase: ____________________

Longitude: ____________________ Latitude: ____________________

Object(s) Viewed: ____

Equipment:

Filter	Magnification	EP	Aperture	FOV

Conditions	Seeing	Finder

Transparency: ____

Notes:

Date:________________________ Time: ________________________

Location:________________________ Lunar Phase: ________________________

Longitude: ________________________ Latitude: ________________________

Object(s) Viewed: ____

Equipment:

Filter	Magnification	EP	Aperture	FOV

Conditions	Seeing	Finder

Transparency: ____

Notes:

Date:________________________ Time: ________________________

Location:______________________ Lunar Phase: ____________________

Longitude: ____________________ Latitude: ______________________

Object(s) Viewed: ____

Equipment:

Filter	Magnification	EP	Aperture	FOV

Conditions	Seeing	Finder

Transparency: ____

Notes:

Date: ______________________________ Time: ______________________________

Location: ___________________________ Lunar Phase: ________________________

Longitude: __________________________ Latitude: ___________________________

Object(s) Viewed: ____

Equipment:

Filter	Magnification	EP	Aperture	FOV

Conditions	Seeing	Finder

Transparency: ____

Notes:

Date: ______________________ Time: ______________________

Location: ______________________ Lunar Phase: ______________________

Longitude: ______________________ Latitude: ______________________

Object(s) Viewed: ____

Equipment:

Filter	Magnification	EP	Aperture	FOV

Conditions	Seeing	Finder

Transparency: ____

Notes:

Date:____________________________ Time: _______________________________

Location:_________________________ Lunar Phase: _________________________

Longitude: _______________________ Latitude: ____________________________

Object(s) Viewed: ____

Equipment:

Filter	Magnification	EP	Aperture	FOV

Conditions	Seeing	Finder

Transparency: ____

Notes:

Date:________________ Time: ________________

Location:________________ Lunar Phase: ________________

Longitude: ________________ Latitude: ________________

Object(s) Viewed: ________________________________

Equipment:

Filter	Magnification	EP	Aperture	FOV

Conditions	Seeing	Finder

Transparency: ________________________________

Notes:

Date: ____________________ Time: ____________________

Location: ____________________ Lunar Phase: ____________________

Longitude: ____________________ Latitude: ____________________

Object(s) Viewed: ____________________

Equipment:

Filter	Magnification	EP	Aperture	FOV

Conditions	Seeing	Finder

Transparency: ____________________

Notes:

Date:________________________ Time: ________________________

Location:______________________ Lunar Phase: ____________________

Longitude: ______________________ Latitude: ______________________

Object(s) Viewed: ____

Equipment:

Filter	Magnification	EP	Aperture	FOV

Conditions	Seeing	Finder

Transparency: ____

Notes:

Date: ____________________ Time: ____________________

Location: ____________________ Lunar Phase: ____________________

Longitude: ____________________ Latitude: ____________________

Object(s) Viewed: ____

Equipment:

Filter	Magnification	EP	Aperture	FOV

Conditions	Seeing	Finder

Transparency: ____

Notes:

Date: ______________________ Time: ______________________

Location: ______________________ Lunar Phase: ______________________

Longitude: ______________________ Latitude: ______________________

Object(s) Viewed: ____

____

Equipment:

Filter	Magnification	EP	Aperture	FOV

Conditions	Seeing	Finder

Transparency: ____

	Notes:

Date:____________________ Time: ____________________

Location:____________________ Lunar Phase: ____________________

Longitude: ____________________ Latitude: ____________________

Object(s) Viewed: ____

Equipment:

Filter	Magnification	EP	Aperture	FOV

Conditions	Seeing	Finder

Transparency: ____

Notes:

Date: ______________________ Time: ______________________

Location: ______________________ Lunar Phase: ______________________

Longitude: ______________________ Latitude: ______________________

Object(s) Viewed: ____

Equipment:

Filter	Magnification	EP	Aperture	FOV

Conditions	Seeing	Finder

Transparency: ____

Notes:

Date: ____________________ Time: ____________________

Location: ____________________ Lunar Phase: ____________________

Longitude: ____________________ Latitude: ____________________

Object(s) Viewed: ____

Equipment:

Filter	Magnification	EP	Aperture	FOV

Conditions	Seeing	Finder

Transparency: ____

Notes:

Date:___________________________ Time: ______________________________

Location:__________________________ Lunar Phase: _________________________

Longitude: ______________________ Latitude: ___________________________

Object(s) Viewed: ____

Equipment:

Filter	Magnification	EP	Aperture	FOV

Conditions	Seeing	Finder

Transparency: ____

Notes:

Date:______________________________ Time: ______________________________

Location:________________________ Lunar Phase: ________________________

Longitude: ______________________ Latitude: ___________________________

Object(s) Viewed: ____

Equipment:

Filter	Magnification	EP	Aperture	FOV

Conditions	Seeing	Finder

Transparency: ____

Notes:

Date: ____________________ Time: ____________________

Location: ____________________ Lunar Phase: ____________________

Longitude: ____________________ Latitude: ____________________

Object(s) Viewed: ____

Equipment:

Filter	Magnification	EP	Aperture	FOV

Conditions	Seeing	Finder

Transparency: ____

Notes:

Date:____________________ Time: ____________________

Location:____________________ Lunar Phase: ____________________

Longitude: ____________________ Latitude: ____________________

Object(s) Viewed: ____

Equipment:

Filter	Magnification	EP	Aperture	FOV

Conditions	Seeing	Finder

Transparency: ____

Notes:

Date: ______________________________ Time: ______________________________

Location: ____________________________ Lunar Phase: ________________________

Longitude: ___________________________ Latitude: ____________________________

Object(s) Viewed: ____

Equipment:

Filter	Magnification	EP	Aperture	FOV

Conditions	Seeing	Finder

Transparency: ____

Notes:

Date:______________________ Time: ______________________

Location:______________________ Lunar Phase: ______________________

Longitude: ______________________ Latitude: ______________________

Object(s) Viewed: ____

Equipment:

Filter	Magnification	EP	Aperture	FOV

Conditions	Seeing	Finder

Transparency: ____

Notes:

Date:____________________ Time: ____________________
Location:____________________ Lunar Phase: ____________________
Longitude: ____________________ Latitude: ____________________
Object(s) Viewed: ____________________

Equipment:

Filter	Magnification	EP	Aperture	FOV

Conditions	Seeing	Finder

Transparency: ____________________

Notes:

Date:_______________________ Time: _______________________

Location:_______________________ Lunar Phase: _______________________

Longitude: _______________________ Latitude: _______________________

Object(s) Viewed: ____

Equipment:

Filter	Magnification	EP	Aperture	FOV

Conditions	Seeing	Finder

Transparency: ____

Notes:

Date:______________________________ Time: ______________________________

Location:___________________________ Lunar Phase: ________________________

Longitude: _________________________ Latitude: ___________________________

Object(s) Viewed: ____

Equipment:

Filter	Magnification	EP	Aperture	FOV

Conditions	Seeing	Finder

Transparency: ____

Notes:

Date:____________________ Time: ____________________

Location:____________________ Lunar Phase: ____________________

Longitude: ____________________ Latitude: ____________________

Object(s) Viewed: ____

Equipment:

Filter	Magnification	EP	Aperture	FOV

Conditions	Seeing	Finder

Transparency: ____

Notes:

Date: ______________________ Time: ______________________

Location: ______________________ Lunar Phase: ______________________

Longitude: ______________________ Latitude: ______________________

Object(s) Viewed: ______________________

____

Equipment:

Filter	Magnification	EP	Aperture	FOV

Conditions	Seeing	Finder

Transparency: ______________________

Notes:

Date:______________________________ Time: ________________________________

Location:__________________________ Lunar Phase: __________________________

Longitude: ________________________ Latitude: _____________________________

Object(s) Viewed: ____

Equipment:

Filter	Magnification	EP	Aperture	FOV

Conditions	Seeing	Finder

Transparency: ____

Notes:

Date: ______________________________ Time: ______________________________

Location: ______________________________ Lunar Phase: ______________________________

Longitude: ______________________________ Latitude: ______________________________

Object(s) Viewed: ____

Equipment:

Filter	Magnification	EP	Aperture	FOV

Conditions	Seeing	Finder

Transparency: ____

Notes:

Date:____________________ Time: ____________________

Location:____________________ Lunar Phase: ____________________

Longitude: ____________________ Latitude: ____________________

Object(s) Viewed: ____

Equipment:

Filter	Magnification	EP	Aperture	FOV

Conditions	Seeing	Finder

Transparency: ____

Notes:

Date: ______________________ Time: ______________________

Location: ______________________ Lunar Phase: ______________________

Longitude: ______________________ Latitude: ______________________

Object(s) Viewed: ____

Equipment:

Filter	Magnification	EP	Aperture	FOV

Conditions	Seeing	Finder

Transparency: ____

Notes:

Date:______________________ Time: ______________________

Location:______________________ Lunar Phase: ______________________

Longitude: ______________________ Latitude: ______________________

Object(s) Viewed: ____

Equipment:

Filter	Magnification	EP	Aperture	FOV

Conditions	Seeing	Finder

Transparency: ____

Notes:

Date: ____________________ Time: ____________________

Location: ____________________ Lunar Phase: ____________________

Longitude: ____________________ Latitude: ____________________

Object(s) Viewed: ____

Equipment:

Filter	Magnification	EP	Aperture	FOV

Conditions	Seeing	Finder

Transparency: ____

	Notes:

Date:______________________________ Time: ________________________________

Location:__________________________ Lunar Phase: __________________________

Longitude: ________________________ Latitude: _____________________________

Object(s) Viewed: ____

Equipment:

Filter	Magnification	EP	Aperture	FOV

Conditions	Seeing	Finder

Transparency: ____

Notes:

Date:________________________ Time: ________________________

Location:____________________ Lunar Phase: ___________________

Longitude: __________________ Latitude: ______________________

Object(s) Viewed: ____

Equipment:

Filter	Magnification	EP	Aperture	FOV

Conditions	Seeing	Finder

Transparency: ____

	Notes:

Date: ____________________ Time: ____________________

Location: ____________________ Lunar Phase: ____________________

Longitude: ____________________ Latitude: ____________________

Object(s) Viewed: ____

Equipment:

Filter	Magnification	EP	Aperture	FOV

Conditions	Seeing	Finder

Transparency: ____

Notes:

Date: ______________________________ Time: ______________________________

Location: ______________________________ Lunar Phase: ______________________________

Longitude: ______________________________ Latitude: ______________________________

Object(s) Viewed: ____

Equipment:

Filter	Magnification	EP	Aperture	FOV

Conditions	Seeing	Finder

Transparency: ____

Notes:

Date: ______________________ Time: ______________________

Location: ______________________ Lunar Phase: ______________________

Longitude: ______________________ Latitude: ______________________

Object(s) Viewed: ______________________

Equipment:

Filter	Magnification	EP	Aperture	FOV

Conditions	Seeing	Finder

Transparency: ______________________

Notes:

Date: ______________________ Time: ______________________

Location: ______________________ Lunar Phase: ______________________

Longitude: ______________________ Latitude: ______________________

Object(s) Viewed: ____

Equipment:

Filter	Magnification	EP	Aperture	FOV

Conditions	Seeing	Finder

Transparency: ____

Notes:

Date: ____________________ Time: ____________________

Location: ____________________ Lunar Phase: ____________________

Longitude: ____________________ Latitude: ____________________

Object(s) Viewed: ____

Equipment:

Filter	Magnification	EP	Aperture	FOV

Conditions	Seeing	Finder

Transparency: ____

Notes:

Date: ______________________________ Time: ______________________________

Location: ______________________________ Lunar Phase: ______________________________

Longitude: ______________________________ Latitude: ______________________________

Object(s) Viewed: ______________________________

Equipment:

Filter	Magnification	EP	Aperture	FOV

Conditions	Seeing	Finder

Transparency: ______________________________

Notes:

Date: ______________________ Time: ______________________

Location: ______________________ Lunar Phase: ______________________

Longitude: ______________________ Latitude: ______________________

Object(s) Viewed: ____

Equipment:

Filter	Magnification	EP	Aperture	FOV

Conditions	Seeing	Finder

Transparency: ____

Notes:

Date:________________________ Time: ________________________

Location:______________________ Lunar Phase: ____________________

Longitude: ____________________ Latitude: _______________________

Object(s) Viewed: ____

Equipment:

Filter	Magnification	EP	Aperture	FOV

Conditions	Seeing	Finder

Transparency: ____

Notes:

Date:______________________________ Time: ________________________________

Location:__________________________ Lunar Phase: __________________________

Longitude: ________________________ Latitude: _____________________________

Object(s) Viewed: ____

Equipment:

Filter	Magnification	EP	Aperture	FOV

Conditions	Seeing	Finder

Transparency: ____

Notes:

Date: ______________________ Time: ______________________

Location: ______________________ Lunar Phase: ______________________

Longitude: ______________________ Latitude: ______________________

Object(s) Viewed: ____

Equipment:

Filter	Magnification	EP	Aperture	FOV

Conditions	Seeing	Finder

Transparency: ____

Notes:

Date: ____________________ Time: ____________________

Location: ____________________ Lunar Phase: ____________________

Longitude: ____________________ Latitude: ____________________

Object(s) Viewed: ____

Equipment:

Filter	Magnification	EP	Aperture	FOV

Conditions	Seeing	Finder

Transparency: ____

Notes:

Date:________________________ Time: ________________________

Location:________________________ Lunar Phase: ________________________

Longitude: ________________________ Latitude: ________________________

Object(s) Viewed: ____

Equipment:

Filter	Magnification	EP	Aperture	FOV

Conditions	Seeing	Finder

Transparency: ____

Notes:

Date: ______________________ Time: ______________________

Location: ______________________ Lunar Phase: ______________________

Longitude: ______________________ Latitude: ______________________

Object(s) Viewed: ____

Equipment:

Filter	Magnification	EP	Aperture	FOV

Conditions	Seeing	Finder

Transparency: ____

Notes:

Date:____________________ Time: ____________________

Location:____________________ Lunar Phase: ____________________

Longitude: ____________________ Latitude: ____________________

Object(s) Viewed: ____

Equipment:

Filter	Magnification	EP	Aperture	FOV

Conditions	Seeing	Finder

Transparency: ____

Notes:

Made in the USA
Monee, IL
03 May 2026

49418503R00057